MIMICRY, AND OTHER PROTECTIVE RESEMBLANCES AMONG ANIMALS

BY

ALFRED RUSSEL WALLACE

British Library Cataloguing-in-Publication Data
A catalogue record for this book is available from the
British Library

Contents

4

Alfred Russel Wallace

Alfred Russel Wallace was born on 8th January 1823 in the village of Llanbadoc, in Monmouthshire, Wales.

At the age of five, Wallace's family moved to Hertford where he later enrolled at Hertford Grammar School. He was educated there until financial difficulties forced his family to withdraw him in 1836. He then boarded with his older brother John before becoming an apprentice to his eldest brother, William, a surveyor. He worked for William for six years until the business declined due to difficult economic conditions.

After a brief period of unemployment, he was hired as a master at the Collegiate School in Leicester to teach drawing, map-making, and surveying. During this time he met the entomologist Henry Bates who inspired Wallace to begin collecting insects. He and bates continued exchanging letters after Wallace left teaching to pursue his surveying career. They corresponded on prominent works of the time such as Charles Darwin's *The Voyage of the Beagle* (1839) and Robert Chamber's *Vestiges of the Natural History of Creation* (1844).

Wallace was inspired by the travelling naturalists of the day and decided to begin his exploration career collecting specimens in the Amazon rainforest. He explored the Rio Negra for four years, making notes on the peoples and languages he encountered as well as the geography, flora, and fauna. On his return voyage his ship, Helen, caught fire and he and the crew were stranded for ten days before being picked up by the Jordeson, a brig travelling from Cuba to London. All of his specimens aboard Helen had been lost.

After a brief stay in England he embarked on a journey to the Malay Archipelago (now Singapore, Malaysia, and Indonesia). During this eight year period he collected more than 126,000 specimens, several thousand of which represented new species to science. While travelling, Wallace refined his thoughts about evolution and in 1858 he outlined his theory of natural selection in an article he sent to Charles Darwin. This was published in the same year along with Darwin's own theory. Wallace eventually published an account of his travels *The Malay Archipelago* in 1869, and it became one of the most popular books of scientific exploration in the 19th century.

Upon his return to England, in 1862, Wallace became a staunch defender of Darwin's landmark work *On the Origin of Species* (1859). He wrote responses to those critical of the theory of natural selection, including 'Remarks on the Rev. S. Haughton's Paper on the Bee's Cell, And on the Origin of Species' (1863) and 'Creation by Law' (1867). The former of these was particularly pleasing to Darwin. Wallace also published important papers such as 'The Origin of Human Races and the Antiquity of Man Deduced from the Theory of 'Natural Selection'' (1864) and books, including the much cited *Darwinism* (1889).

Wallace made a huge contribution to the natural sciences and he will continue to be remembered as one of the key figures in the development of evolutionary theory.

Wallace died on 7th November 1913 at the age of 90. He is buried in a small cemetery at Broadstone, Dorset, England.

MIMICRY, AND OTHER PROTECTIVE RESEMBLANCES AMONG ANIMALS

There is no more convincing proof of the truth of a comprehensive theory, than its power of absorbing and finding a place for new facts, and its capability of interpreting phænomena which had been previously looked upon as unaccountable anomalies. It is thus that the law of universal gravitation and the undulatory theory of light have become established and universally accepted by men of science. Fact after fact has been brought forward as being apparently inconsistent with them, and one after another these very facts have been shown to be the consequences of the laws they were at first supposed to disprove. A false theory will never stand this test. Advancing knowledge brings to light whole groups of facts which it cannot deal with, and its advocates steadily decrease in numbers, notwithstanding the ability and scientific skill with which it may have been supported. The great name of Edward Forbes did not prevent his theory of "Polarity in the distribution of Organic beings in Time" from dying a natural death; but the most striking illustration of the behaviour of a false theory is to be found in the "Circular and Quinarian System" of classification propounded by MacLeay, and developed by Swainson, with an amount of knowledge and ingenuity that have rarely been surpassed. This theory was eminently attractive, both from its symmetry and completeness, and from the interesting nature of the varied analogies and affinities which it brought to light and made use of. The series of Natural History volumes in "Lardner's Cabinet Cyclopædia," in which Mr. Swainson developed it in most departments of the animal kingdom, made it widely known; and in fact for a long time these were the best and almost the only popular text-books for the rising generation of naturalists. It was favourably received too by the older school, which was perhaps rather an indication of its unsoundness. A considerable number of well-known naturalists either spoke approvingly of it, or advocated similar principles, and for a good many years it was decidedly in the ascendent. With such a favourable introduction, and with such talented exponents, it must have become established if it had had any germ of truth in it; yet it quite died out in a few short years, its very existence is now a matter of history; and so rapid was its fall that its talented creator, Swainson, perhaps lived to be the last man who believed in it.

Such is the course of a false theory. That of a true one is very different, as may be well seen by the progress of opinion on the subject of Natural Selection. In less than eight years "The Origin of Species" has produced conviction in the minds of a majority of the most eminent living men of science. New facts, new problems, new difficulties as they arise are accepted, solved or removed by this theory; and its principles are illustrated by the progress and conclusions of every well established branch of human knowledge. It is the object of the present essay to show how it has recently been applied to connect together and explain a variety of curious facts which had long been considered as inexplicable anomalies.

Importance of the Principle of Utility.

Perhaps no principle has ever been announced so fertile in results as that which Mr. Darwin so earnestly impresses upon us, and which is indeed a necessary deduction from the theory of Natural Selection, namely — that none of the definite facts of organic nature, no special organ, no characteristic form or marking, no peculiarities of instinct or of habit, no relations between species or between groups of species — can exist, but which must now be or once have been *useful* to the individuals or the races which possess them. This great principle gives us a clue which we can follow out in the study of many recondite phænomena, and leads us to seek a meaning and a purpose of some definite character in minutiæ which we should be otherwise almost sure to pass over as insignificant or unimportant.

Popular Theories of Colour in Animals.

The adaptation of the external colouring of animals to their conditions of life has long been recognised, and has been imputed either to an originally created specific peculiarity, or to the direct action of climate, soil, or food. Where the former explanation has been accepted, it has completely checked inquiry, since we could never get any further than the fact of the adaptation. There was nothing more to be known about the matter. The second explanation was soon found to be quite inadequate to deal with all the varied phases of the phænomena, and to be contradicted by many well-known facts. For example, wild rabbits are always of grey or brown tints well suited for concealment among grass and fern. But when these rabbits are domesticated, without any change of climate or food, they vary into white or black, and these varieties may be multiplied to any extent, forming white or black races. Exactly the same thing has occurred with pigeons; and in the case of rats and mice, the white variety has not been shown to be at all dependent on alteration of climate, food, or other external conditions. In many cases the wings of an insect not only assume the exact tint of the bark or leaf it is accustomed to rest on, but the form and veining of the leaf or the exact rugosity of the bark is imitated; and these detailed modifications cannot be reasonably imputed to climate or to food, since in many cases the species does not feed on the substance it resembles, and when it does, no reasonable connexion can be shown to exist between the supposed cause and the effect produced. It was reserved for the theory of Natural Selection to solve all these problems, and many others which were not at first supposed to be directly connected with them. To make these latter intelligible, it will be necessary to give a sketch of the whole series of phænomena which may be classed under the head of useful or protective resemblances.

Importance of Concealment as Influencing Colour.

Concealment, more or less complete, is useful to many animals, and absolutely essential to some. Those which have numerous enemies from which they cannot escape by rapidity of motion, find safety in concealment. Those which prey upon others must also be so constituted as not to alarm them by their presence or their approach, or they would soon die of hunger. Now it is remarkable in how many cases nature gives this boon to the animal, by colouring it with such tints as may best serve to enable it to escape from its enemies or to entrap its prey. Desert animals as a rule are desert-coloured. The lion is a typical example of this, and must be almost invisible when crouched upon the sand or among desert rocks and stones. Antelopes are all more or less sandy-coloured. The camel is preeminently so. The Egyptian cat and the Pampas cat are sandy or earth-coloured. The Australian kangaroos are of the same tints, and the original colour of the wild horse is supposed to have been a sandy or clay-colour.

The desert birds are still more remarkably protected by their assimilative hues. The stonechats, the larks, the quails, the goatsuckers and the grouse, which abound in the North African and Asiatic deserts, are all tinted and mottled so as to resemble with wonderful accuracy the average colour and aspect of the soil in the district they inhabit. The Rev. H. Tristram, in his account of the ornithology of North Africa in the 1st volume of the "Ibis," says: "In the desert, where neither trees, brush-wood, nor even undulation of the surface afford the slightest protection to its foes, a modification of colour which shall be assimilated to that of the surrounding country, is absolutely necessary. Hence *without exception* the upper plumage of *every bird*, whether lark, chat, sylvain, or sand-grouse, and also the fur of *all the smaller mammals*, and the skin of *all the snakes and lizards*, is of one uniform isabelline or sand colour." After the testimony of so able an observer it is unnecessary to adduce further examples of the protective colours of desert animals.

Almost equally striking are the cases of arctic animals possessing the white colour that best conceals them upon snowfields and icebergs. The polar bear is the only bear that is white, and it lives constantly among snow and ice. The arctic fox, the ermine and the alpine hare change to white in winter only, because in summer white would be more conspicuous than any other colour, and therefore a danger rather than a protection; but the American polar hare, inhabiting regions of almost perpetual snow, is white all the year round. Other animals inhabiting the same Northern regions do not, however, change colour. The sable is a good example, for throughout the severity of a Siberian winter it retains its rich brown fur. But its habits are such that it does not need the protection of colour, for it is said to be able to subsist on fruits and berries in winter, and to be so active upon the trees as to catch small birds among the branches. So also the woodchuck of Canada has a dark-brown fur; but then it lives in burrows and frequents river banks, catching fish and small animals that live in or near the water.

Among birds, the ptarmigan is a fine example of protective colouring. Its summer plumage so exactly harmonizes with the lichen-coloured stones among which it delights to sit, that a person may walk through a flock of them without seeing a single bird; while in winter its white plumage is an almost equal protection. The snow-bunting, the jer-falcon, and the snowy owl are also white-coloured birds inhabiting the arctic regions, and there can be little doubt but that their colouring is to some extent protective.

Nocturnal animals supply us with equally good illustrations. Mice, rats, bats, and moles possess the least conspicuous of hues, and must be quite invisible at times when any light colour would be instantly seen. Owls and goatsuckers are of those dark mottled tints that will assimilate with bark and lichen, and thus protect them during the day, and at the same time be inconspicuous in the dusk.

It is only in the tropics, among forests which never lose their foliage, that we find whole groups of birds whose chief colour is green. The parrots are the most striking example, but we have also a group of green pigeons in the East; and the barbets, leaf-thrushes, bee-eaters, white-eyes, turacos, and several smaller groups, have so much green in their plumage as to tend greatly to conceal them among the foliage.

Special Modifications of Colour.

The conformity of tint which has been so far shown to exist between animals and their habitations is of a somewhat general character; we will now consider the cases of more special adaptation. If the lion is enabled by his sandy colour readily to conceal himself by merely crouching down upon the desert, how, it may be asked, do the elegant markings of the tiger, the jaguar, and the other large cats agree with this theory? We reply that these are generally cases of more or less special adaptation. The tiger is a jungle animal, and hides himself among tufts of grass or of bamboos, and in these positions the vertical stripes with which his body is adorned must so assimilate with the vertical stems of the bamboo, as to assist greatly in concealing him from his approaching prey. How remarkable it is that besides the lion and tiger, almost all the other large cats are arboreal in their habits, and almost all have ocellated or spotted skins, which must certainly tend to blend them with the background of foliage; while the one exception, the puma, has an ashy brown uniform fur, and has the habit of clinging so closely to a limb of a tree while waiting for his prey to pass beneath as to be hardly distinguishable from the bark.

Among birds, the ptarmigan, already mentioned, must be considered a remarkable case of special adaptation. Another is a South–American goatsucker (Caprimulgus rupestris) which rests in the bright sunshine on little bare rocky islets in the Upper Rio Negro, where its unusually light colours so closely resemble those of the rock and sand, that it can scarcely be detected till trodden upon.

The Duke of Argyll, in his "Reign of Law," has pointed out the admirable adaptation of the colours of the woodcock to its protection. The various browns and yellows and pale ash-colour that occur in fallen leaves are all reproduced in its plumage, so that when according to its habit it rests upon the ground under trees, it is almost impossible to detect it. In snipes the colours are modified so as to be equally in harmony with the prevalent forms and colours of marshy vegetation. Mr. J. M. Lester, in a paper read before the Rugby School Natural History Society, observes:—"The wood-dove, when perched amongst the branches of its favourite *fir*, is scarcely discernible; whereas, were it among some lighter foliage, the blue and purple tints in its plumage would far sooner betray it. The robin redbreast too, although it might be thought that the red on its breast made it much easier to be seen, is in reality not at all endangered by it, since it generally contrives to get among some russet or yellow fading leaves, where the red matches very well with the autumn tints, and the brown of the rest of the body with the bare branches."

Reptiles offer us many similar examples. The most arboreal lizards, the iguanas, are as green as the leaves they feed upon, and the slender whip-snakes are rendered almost invisible as they glide among the foliage by a similar colouration. How difficult it is sometimes to catch sight of the little green tree-frogs sitting on the leaves of a small plant enclosed in a glass case in the Zoological Gardens; yet how much better concealed must they be among the fresh green damp foliage of a marshy forest. There is a North–American frog found on lichen-covered rocks and walls, which is so coloured as exactly to resemble them, and as long as it remains quiet would certainly escape detection. Some of the geckos which cling motionless on the trunks of trees in the tropics, are of such curiously marbled colours as to match exactly with the bark they rest upon.

In every part of the tropics there are tree-snakes that twist among boughs and shrubs, or lie coiled up on the dense masses of foliage. These are of many distinct groups, and comprise both venomous and harmless genera; but almost all of them are of a beautiful green colour, sometimes more or less adorned with white or dusky bands and spots. There can be little doubt that this colour is doubly useful to them, since it will tend to conceal them from their enemies, and will lead their prey to approach them unconscious of danger. Dr. Gunther informs me that there is only one genus of true arboreal snakes (Dipsas) whose colours are rarely green, but are of various shades of black, brown, and olive, and these are all nocturnal reptiles, and there can be little doubt conceal themselves during the day in holes, so that the green protective tint would be useless to them, and they accordingly retain the more usual reptilian hues.

Fishes present similar instances. Many flat fish, as for example the flounder and the skate, are exactly the colour of the gravel or sand on which they habitually rest. Among the marine flower gardens of an Eastern coral reef the fishes present every variety of gorgeous colour, while the river fish even of the tropics rarely if ever have gay or conspicuous markings. A very curious case of this kind of adaptation occurs in the sea-horses (Hippocampus) of Australia, some of which bear long foliaceous appendages resembling seaweed, and are of a brilliant red colour; and they are known to live among seaweed of the same hue, so that when at rest they must be quite invisible. There are now in the aquarium of the Zoological Society some slender green pipe-fish which fasten themselves to any object at the bottom by their prehensile tails, and float about with the current, looking exactly like some simple cylindrical algæ.

It is, however, in the insect world that this principle of the adaptation of animals to their environment is most fully and strikingly developed. In order to understand how general this is, it is necessary to enter somewhat into details, as we shall thereby be better able to appreciate the significance of the still more remarkable phenomena we shall presently have to discuss. It seems to be in proportion to their sluggish motions or the absence of other means of defence, that insects possess the protective colouring. In the tropics there are thousands of species of insects which rest during the day clinging to the bark of dead or fallen trees; and the greater portion of these are delicately mottled with gray and brown tints, which though symmetrically disposed and infinitely varied, yet blend so completely with the usual colours of the bark, that at two or three feet distance they are quite undistinguishable. In some cases a species is known to frequent only one species of tree. This is the case with the common South American long-horned beetle (Onychocerus scorpio) which, Mr. Bates informed me, is found only on a rough-barked tree, called Tapiribá, on the Amazon. It is very abundant, but so exactly does it resemble the bark in colour and rugosity, and so closely does it cling to the branches, that until it moves it is absolutely invisible! An allied species (O. concentricus) is found only at Pará, on a distinct species of tree, the bark of which it resembles with equal accuracy. Both these insects are abundant, and we may fairly conclude that the protection they derive from this strange concealment is at least one of the causes that enable the race to flourish.

Many of the species of Cicindela, or tiger beetle, will illustrate this mode of protection. Our common Cicindela campestris frequents grassy banks, and is of a beautiful green colour, while C. maritima, which is found only on sandy sea-shores, is of a pale bronzy yellow, so as to be almost invisible. A great number of the species found by myself in the Malay islands are similarly protected. The beautiful Cicindela gloriosa, of a very deep velvety green colour, was only taken upon wet mossy stones in the bed of a mountain stream, where it was with the greatest difficulty detected. A large brown species (C. heros) was found chiefly on dead leaves in forest paths; and one which was never seen except on the wet mud of salt marshes was of a glossy olive so exactly the colour of the mud as only to be distinguished when the sun shone, by its shadow! Where the sandy beach was coralline and nearly white, I found a very pale Cicindela; wherever it was volcanic and black, a dark species of the same genus was sure to be met with.

There are in the East small beetles of the family Buprestidæ which generally rest on the midrib of a leaf, and the naturalist often hesitates before picking them off, so closely do they resemble pieces of bird's dung. Kirby and Spence mention the small beetle Onthophilus sulcatus as being like the seed of an umbelliferous plant; and another small weevil, which is much persecuted by predatory beetles of the genus Harpalus, is of the exact colour of loamy soil, and was found to be particularly abundant in loam pits. Mr. Bates mentions a small beetle (Chlamys pilula) which was undistinguishable by the eye from the dung of caterpillars, while some of the Cassidæ, from their hemispherical forms and pearly gold colour, resemble glittering dew-drops upon the leaves.

A number of our small brown and speckled weevils at the approach of any object roll off the leaf they are sitting on, at the same time drawing in their legs and antennæ, which fit so perfectly into cavities for their reception that the insect becomes a mere oval brownish lump, which it is hopeless to look for among the similarly coloured little stones and earth pellets among which it lies motionless.

The distribution of colour in butterflies and moths respectively is very instructive from this point of view. The former have all their brilliant colouring on the upper surface of all four wings, while the under surface is almost always soberly coloured, and often very dark and obscure. The moths on the contrary have generally their chief colour on the hind wings only, the upper wings being of dull, sombre, and often imitative tints, and these generally conceal the hind wings when the insects are in repose. This arrangement of the colours is therefore eminently protective, because the butterfly always rests with his wings raised so as to conceal the dangerous brilliancy of his upper surface. It is probable that if we watched their habits sufficiently we should find the under surface of the wings of butterflies very frequently imitative and protective. Mr. T. W. Wood has pointed out that the little orange-tip butterfly often rests in the evening on the green and white flower heads of an umbelliferous plant, and that when observed in this position the beautiful green and white mottling of the under surface completely assimilates with the flower heads and renders the creature very difficult to be seen. It is probable that the rich dark colouring of the under side of our peacock, tortoiseshell, and red-admiral butterflies answers a similar purpose.

Two curious South American butterflies that always settle on the trunks of trees (Gynecia dirce and Callizona acesta) have the under surface curiously striped and mottled, and when viewed obliquely must closely assimilate with the appearance of the furrowed bark of many kinds of trees. But the most wonderful and undoubted case of protective resemblance in a butterfly which I have ever seen, is that of the common Indian Kallima inachis, and its Malayan ally, Kallima paralekta. The upper surface of these insects is very striking and showy, as they are of a large size, and are adorned with a broad band of rich orange on a deep bluish ground. The under side is very variable in colour, so that out of fifty specimens no two can be found exactly alike, but every one of them will be of some shade of ash or brown or ochre, such as are found among dead, dry, or decaying leaves. The apex of the upper wings is produced into an acute point, a very common form in the leaves of tropical shrubs and trees, and the lower wings are also produced into a short narrow tail. Between these two points runs a dark curved line exactly representing the midrib of a leaf, and from this radiate on each side a few oblique lines, which serve to indicate the lateral veins of a leaf. These marks are more clearly seen on the outer portion of the base of the wings, and on the inner side towards the middle and apex, and it is very curious to observe how the usual marginal and transverse striæ of the group are here modified and strengthened so as to become adapted for an imitation of the venation of a leaf. We come now to a still more extraordinary part of the imitation, for we find representations of leaves in every stage of decay, variously blotched and mildewed and pierced with holes, and in many cases irregularly covered with powdery black dots gathered into patches and spots, so closely resembling the various kinds of minute fungi that grow on dead leaves that it is impossible to avoid thinking at first sight that the butterflies themselves have been attacked by real fungi.

But this resemblance, close as it is, would be of little use if the habits of the insect did not accord with it. If the butterfly sat upon leaves or upon flowers, or opened its wings so as to expose the upper surface, or exposed and moved its head and antennæ as many other butterflies do, its disguise would be of little avail. We might be sure, however, from the analogy of many other cases, that the habits of the insect are such as still further to aid its deceptive garb; but we are not obliged to make any such supposition, since I myself had the good fortune to observe scores of Kallima paralekta, in Sumatra, and to capture many of them, and can vouch for the accuracy of the following details. These butterflies frequent dry forests and fly very swiftly. They were never seen to settle on a flower or a green leaf, but were many times lost sight of in a bush or tree of dead leaves. On such occasions they were generally searched for in vain, for while gazing intently at the very spot where one had disappeared, it would often suddenly dart out, and again vanish twenty or fifty yards further on. On one or two occasions the insect was detected reposing, and it could then be seen how completely it assimilates itself to the surrounding leaves. It sits on a nearly upright twig, the wings fitting closely back to back, concealing the antennæ and head, which are drawn up between their bases. The little tails of the hind wing touch the branch, and form a perfect stalk to the leaf, which is supported in its place by the claws of the middle pair of feet, which are slender and inconspicuous. The irregular outline of the wings gives exactly the perspective effect of a shrivelled leaf. We thus have size, colour, form, markings, and habits, all combining together to produce a disguise which may be said to be absolutely perfect; and the protection which it affords is sufficiently indicated by the abundance of the individuals that possess it.

The Rev. Joseph Greene has called attention to the striking harmony between the colours of those British moths which are on the wing in autumn and winter, and the prevailing tints of nature at those seasons. In autumn various shades of yellow and brown prevail, and he shows that out of fifty-two species that fly at this season, no less than forty-two are of corresponding colours. Orgyia antiqua, O. gonostigma, the genera Xanthia, Glæa, and Ennomos are examples. In winter, gray and silvery tints prevail, and the genus Chematobia and several species of Hybernia which fly during this season are of corresponding hues. No doubt if the habits of moths in a state of nature were more closely observed, we should find many cases of special protective resemblance. A few such have already been noticed. Agriopis aprilina, Acronycta psi, and many other moths which rest during the day on the north side of the trunks of trees can with difficulty be distinguished from the grey and green lichens that cover them. The lappet moth (Gastropacha querci) closely resembles both in shape and colour a brown dry leaf; and the well-known buff-tip moth, when at rest is like the broken end of a lichen-covered branch. There are some of the small moths which exactly resemble the dung of birds dropped on leaves, and on this point Mr. A. Sidgwick, in a paper read before the Rugby School Natural History Society, gives the following original observation:—"I myself have more than once mistaken Cilix compressa, a little white and grey moth, for a piece of bird's dung dropped upon a leaf, and *vice versâ* the dung for the moth. Bryophila Glandifera and Perla are the very image of the mortar walls on which they rest; and only this summer, in Switzerland, I amused myself for some time in watching a moth, probably Larentia tripunctaria, fluttering about quite close to me, and then alighting on a wall of the stone of the district which it so exactly matched as to be quite invisible a couple of yards off." There are probably hosts of these resemblances which have not been observed, owing to the difficulty of finding many of the species in their stations of natural repose. Caterpillars are also similarly protected. Many exactly resemble in tint the leaves they feed upon; others are like little brown twigs, and many are so strangely marked or humped, that when motionless they can hardly be taken to be living creatures at all. Mr. Andrew Murray has remarked how closely the larva of the peacock moth (Saturnia pavonia-minor) harmonizes in its ground colour with that of the young buds of heather on which it feeds, and that the pink spots with which it is decorated correspond with the flowers and flower-buds of the same plant.

The whole order of Orthoptera, grasshoppers, locusts, crickets, &c., are protected by their colours harmonizing with that of the vegetation or the soil on which they live, and in no other group have we such striking examples of special resemblance. Most of the tropical Mantidæ and Locustidæ are of the exact tint of the leaves on which they habitually repose, and many of them in addition have the veinings of their wings modified so as exactly to imitate that of a leaf. This is carried to the furthest possible extent in the wonderful genus, Phyllium, the "walking leaf," in which not only are the wings perfect imitations of leaves in every detail, but the thorax and legs are flat, dilated, and leaf-like; so that when tho living insect is resting among the foliage on which it feeds, the closest observation is often unable to distinguish between the animal and the vegetable.

The whole family of the Phasmidæ, or spectres, to which this insect belongs, is more or less imitative, and a great number of the species are called "walking-stick insects," from their singular resemblance to twigs and branches. Some of these are a foot long and as thick as one's finger, and their whole colouring, form, rugosity, and the arrangement of the head, legs, and antennæ, are such as to render them absolutely identical in appearance with dead sticks. They hang loosely about shrubs in the forest, and have the extraordinary habit of stretching out their legs unsymmetrically, so as to render the deception more complete. One of these creatures obtained by myself in Borneo (Ceroxylus laceratus) was covered over with foliaceous excrescences of a clear olive green colour, so as exactly to resemble a stick grown over by a creeping moss or jungermannia. The Dyak who brought it me assured me it was grown over with moss although alive, and it was only after a most minute examination that I could convince myself it was not so.

We need not adduce any more examples to show how important are the details of form and of colouring in animals, and that their very existence may often depend upon their being by these means concealed from their enemies. This kind of protection is found apparently in every class and order, for it has been noticed wherever we can obtain sufficient knowledge of the details of an animal's life-history. It varies in degree, from the mere absence of conspicuous colour or a general harmony with the prevailing tints of nature, up to such a minute and detailed resemblance to inorganic or vegetable structures as to realize the talisman of the fairy tale, and to give its possessor the power of rendering itself invisible.

Theory of Protective Colouring.

We will now endeavour to show how these wonderful resemblances have most probably been brought about. Returning to the higher animals, let us consider the remarkable fact of the rarity of white colouring in the mammalia or birds of the temperate or tropical zones in a state of nature. There is not a single white land-bird or quadruped in Europe, except the few arctic or alpine species, to which white is a protective colour. Yet in many of these creatures there seems to be no inherent tendency to avoid white, for directly they are domesticated white varieties arise, and appear to thrive as well as others. We have white mice and rats, white cats, horses, dogs, and cattle, white poultry, pigeons, turkeys, and ducks, and white rabbits. Some of these animals have been domesticated for a long period, others only for a few centuries; but in almost every case in which an animal has been thoroughly domesticated, parti-coloured and white varieties are produced and become permanent.

It is also well known that animals in a state of nature produce white varieties occasionally. Blackbirds, starlings, and crows are occasionally seen white, as well as elephants, deer, tigers, hares, moles, and many other animals; but in no case is a permanent white race produced. Now there are no statistics to show that the normal-coloured parents produce white offspring oftener under domestication than in a state of nature, and we have no right to make such an assumption if the facts can be accounted for without it. But if the colours of animals do really, in the various instances already adduced, serve for their concealment and preservation, then white or any other conspicuous colour must be hurtful, and must in most cases shorten an animal's life. A white rabbit would be more surely the prey of hawk or buzzard, and the white mole, or field mouse, could not long escape from the vigilant owl. So, also, any deviation from those tints best adapted to conceal a carnivorous animal would render the pursuit of its prey much more difficult, would place it at a disadvantage among its fellows, and in a time of scarcity would probably cause it to starve to death. On the other hand, if an animal spreads from a temperate into an arctic district, the conditions are changed. During a large portion of the year, and just when the struggle for existence is most severe, white is the prevailing tint of nature, and dark colours will be the most conspicuous. The white varieties will now have an advantage; they will escape from their enemies or will secure food, while their brown companions will be devoured or will starve; and as "like produces like" is the established rule in nature, the white race will become permanently established, and dark varieties, when they occasionally appear, will soon die out from their want of adaptation to their environment. In each case the fittest will survive, and a race will be eventually produced adapted to the conditions in which it lives.

We have here an illustration of the simple and effectual means by which animals are brought into harmony with the rest of nature. That slight amount of variability in every species, which we often look upon as something accidental or abnormal, or so insignificant as to be hardly worthy of notice, is yet the foundation of all those wonderful and harmonious resemblances which play such an important part in the economy of nature. Variation is generally very small in amount, but it is all that is required, because the change in the external conditions to which an animal is subject is generally very slow and intermittent. When these changes have taken place too rapidly, the result has often been the extinction of species; but the general rule is, that climatal and geological changes go on slowly, and the slight but continual variations in the colour, form, and structure of all animals, has furnished individuals adapted to these changes, and who have become the progenitors of modified races. Rapid multiplication, incessant slight variation, and survival of the fittest — these are the laws which ever keep the organic world in harmony with the inorganic, and with itself. These are the laws which we believe have produced all the cases of protective resemblance already adduced, as well as those still more curious examples we have yet to bring before our readers.

It must always be borne in mind that the more wonderful examples, in which there is not only a general but a special resemblance — as in the walking leaf, the mossy phasma, and the leaf-winged butterfly — represent those few instances in which the process of modification has been going on during an immense series of generations. They all occur in the tropics, where the conditions of existence are the most favourable, and where climatic changes have for long periods been hardly perceptible. In most of them favourable variations both of colour, form, structure, and instinct or habit, must have occurred to produce the perfect adaptation we now behold. All these are known to vary, and favourable variations when not accompanied by others that were unfavourable, would certainly survive. At one time a little step might be made in this direction, at another time in that — a change of conditions might sometimes render useless that which it had taken ages to produce — great and sudden physical modifications might often produce the extinction of a race just as it was approaching perfection, and a hundred checks of which we can know nothing may have retarded the progress towards perfect adaptation; so that we can hardly wonder at there being so few cases in which a completely successful result has been attained as shown by the abundance and wide diffusion of the creatures so protected.

Objection that Colour, as being dangerous, should not exist in Nature.

It is as well here to reply to an objection that will no doubt occur to many readers — that if protection is so useful to all animals, and so easily brought about by variation and survival of the fittest, there ought to be no conspicuously-coloured creatures; and they will perhaps ask how we account for the brilliant birds, and painted snakes, and gorgeous insects, that occur abundantly all over the world. It will be advisable to answer this question rather fully, in order that we may be prepared to understand the phenomena of "mimicry," which it is the special object of this paper to illustrate and explain.

The slightest observation of the life of animals will show us, that they escape from their enemies and obtain their food in an infinite number of ways; and that their varied habits and instincts are in every case adapted to the conditions of their existence. The porcupine and the hedgehog have a defensive armour that saves them from the attacks of most animals. The tortoise is not injured by the conspicuous colours of his shell, because that shell is in most cases an effectual protection to him. The skunks of North America find safety in their power of emitting an unbearably offensive odour; the beaver in its aquatic habits and solidly constructed abode. In some cases the chief danger to an animal occurs at one particular period of its existence, and if that is guarded against its numbers can easily be maintained. This is the case with many birds, the eggs and young of which are especially obnoxious to danger, and we find accordingly a variety of curious contrivances to protect them. We have nests carefully concealed, hung from the slender extremities of grass or boughs over water, or placed in the hollow of a tree with a very small opening. When these precautions are successful, so many more individuals will be reared than can possibly find food during the least favourable seasons, that there will always be a number of weakly and inexperienced young birds who will fall a prey to the enemies of the race, and thus render necessary for the stronger and healthier individuals no other safeguard than their strength and activity. The instincts most favourable to the production and rearing of offspring will in these cases be most important, and the survival of the fittest will act so as to keep up and advance those instincts, while other causes which tend to modify colour and marking may continue their action almost unchecked.

It is perhaps in insects that we may best study the varied means by which animals are defended or concealed. One of the uses of the phosphorescence with which many insects are furnished, is probably to frighten away their enemies; for Kirby and Spence state that a ground beetle (Carabus) has been observed running round and round a luminous centipede as if afraid to attack it. An immense number of insects have stings, and some stingless ants of the genus Polyrachis are armed with strong and sharp spines on the back, which must render them unpalatable to many of the smaller insectivorous birds. Many beetles of the family Curculionidæ have the wing cases and other external parts so excessively hard, that they cannot be pinned without first drilling a hole to receive the pin, and it is probable that all such find a protection in this excessive hardness. Great numbers of insects hide themselves among the petals of flowers, or in the cracks of bark and timber; and finally, extensive groups and even whole orders have a more or less powerful and disgusting smell and taste, which they either possess permanently, or can emit at pleasure. The attitudes of some insects may also protect them, as the habit of turning up the tail by the harmless rove-beetles (Staphylindidæ) no doubt leads other animals besides children to the belief that they can sting. The curious attitude assumed by sphinx caterpillars is probably a safeguard, as well as the blood-red tentacles which can suddenly be thrown out from the neck, by the caterpillars of all the true swallow-tailed butterflies.

It is among the groups that possess some of these varied kinds of protection in a high degree, that we find the greatest amount of conspicuous colour, or at least the most complete absence of protective imitation. The stinging Hymenoptera, wasps, bees, and hornets, are, as a rule, very showy and brilliant insects, and there is not a single instance recorded in which any one of them is coloured so as to resemble a vegetable or inanimate substance. The Chrysididæ, or golden wasps, which do not sting, possess as a substitute the power of rolling themselves up into a ball, which is almost as hard and polished as if really made of metal,— and they are all adorned with the most gorgeous colours. The whole order Hemiptera (comprising the bugs) emit a powerful odour, and they present a very large proportion of gay-coloured and conspicuous insects. The lady-birds (Coccinellidæ) and their allies the Eumorphidæ, are often brightly spotted, as if to attract attention; but they can both emit fluids of a very disagreeable nature, they are certainly rejected by some birds, and are probably never eaten by any.

The great family of ground beetles (Carabidæ) almost all possess a disagreeable and some a very pungent smell, and a few, called bombardier beetles, have the peculiar faculty of emitting a jet of very volatile liquid, which appears like a puff of smoke, and is accompanied by a distinct crepitating explosion. It is probably because these insects are mostly nocturnal and predacious that they do not present more vivid hues. They are chiefly remarkable for brilliant metallic tints or dull red patches when they are not wholly black, and are therefore very conspicuous by day, when insect-eaters are kept off by their bad odour and taste, but are sufficiently invisible at night when it is of importance that their prey should not become aware of their proximity.

It seems probable that in some cases that which would appear at first to be a source of danger to its possessor may really be a means of protection. Many showy and weak-flying butterflies have a very broad expanse of wing, as in the brilliant blue Morphos of Brazilian forests, and the large Eastern Papilios; yet these groups are tolerably plentiful. Now, specimens of these butterflies are often captured with pierced and broken wings, as if they had been seized by birds from whom they had escaped; but if the wings had been much smaller in proportion to the body, it seems probable that the insect would be more frequently struck or pierced in a vital part, and thus the increased expanse of the wings may have been indirectly beneficial.

In other cases the capacity of increase in a species is so great that however many of the perfect insect may be destroyed, there is always ample means for the continuance of the race. Many of the flesh flies, gnats, ants, palm-tree weevils and locusts are in this category. The whole family of Cetoniadæ or rose chafers, so full of gaily-coloured species, are probably saved from attack by a combination of characters. They fly very rapidly with a zigzag or waving course; they hide themselves the moment they alight, either in the corolla of flowers, or in rotten wood, or in cracks and hollows of trees, and they are generally encased in a very hard and polished coat of mail which may render them unsatisfactory food to such birds as would be able to capture them. The causes which lead to the development of colour have been here able to act unchecked, and we see the result in a large variety of the most gorgeously-coloured insects.

Here, then, with our very imperfect knowledge of the life-history of animals, we are able to see that there are widely varied modes by which they may obtain protection from their enemies or concealment from their prey. Some of those seem to be so complete and effectual as to answer all the wants of the race, and lead to the maintenance of the largest possible population. When this is the case, we can well understand that no further protection derived from a modification of colour can be of the slightest use, and the most brilliant hues may be developed without any prejudicial effect upon the species. On some of the laws that determine the development of colour something may be said presently. It is now merely necessary to show that concealment by obscure or imitative tints is only one out of very many ways by which animals maintain their existence; and having done this we are prepared to consider the phenomena of what has been termed "mimicry." It is to be particularly observed, however, that the word is not here used in the sense of voluntary imitation, but to imply a particular kind of resemblance — a resemblance not in internal structure but in external appearance — a resemblance in those parts only that catch the eye — a resemblance that deceives. As this kind of resemblance has the same effect as voluntary imitation or mimicry, and as we have no word that expresses the required meaning, "mimicry" was adopted by Mr. Bates (who was the first to explain the facts), and has led to some misunderstanding; but there need be none, if it is remembered that both "mimicry" and "imitation" are used in a metaphorical sense, as implying that close external likeness which causes things unlike in structure to be mistaken for each other.

Mimicry.

It has been long known to entomologists that certain insects bear a strange external resemblance to others belonging to distinct genera, families, or even orders, and with which they have no real affinity whatever. The fact, however, appears to have been generally considered as dependent upon some unknown law of "analogy"— some "system of nature," or "general plan," which had guided the Creator in designing the myriads of insect forms, and which we could never hope to understand. In only one case does it appear that the resemblance was thought to be useful, and to have been designed as a means to a definite and intelligible purpose. The flies of the genus Volucella enter the nests of bees to deposit their eggs, so that their larvæ may feed upon the larvæ of the bees, and these flies are each wonderfully like the bee on which it is parasitic. Kirby and Spence believed that this resemblance or "mimicry" was for the express purpose of protecting the flies from the attacks of the bees, and the connection is so evident that it was hardly possible to avoid this conclusion. The resemblance, however, of moths to butterflies or to bees, of beetles to wasps, and of locusts to beetles, has been many times noticed by eminent writers; but scarcely ever till within the last few years does it appear to have been considered that these resemblances had any special purpose, or were of any direct benefit to the insects themselves. In this respect they were looked upon as accidental, as instances of the "curious analogies" in nature which must be wondered at but which could not be explained. Recently, however, these instances have been greatly multiplied; the nature of the resemblances has been more carefully studied, and it has been found that they are often carried out into such details as almost to imply a purpose of deceiving the observer. The phenomena, moreover, have been shown to follow certain definite laws, which again all indicate their dependence on the more general law of the "survival of the fittest," or "the preservation of favoured races in the struggle for life." It will, perhaps, be as well here to state what these laws or general conclusions are, and then to give some account of the facts which support them.

The first law is, that in an overwhelming majority of cases of mimicry, the animals (or the groups) which resemble each other inhabit the same country, the same district, and in most cases are to be found together on the very same spot.

The second law is, that these resemblances are not indiscriminate, but are limited to certain groups, which in every case are abundant in species and individuals, and can often be ascertained to have some special protection.

The third law is, that the species which resemble or "mimic" these dominant groups, are comparatively less abundant in individuals, and are often very rare.

These laws will be found to hold good, in all the cases of true mimicry among various classes of animals to which we have now to call the attention of our readers.

Mimicry among Lepidoptera.

As it is among butterflies that instances of mimicry are most numerous and most striking, an account of some of the more prominent examples in this group will first be given. There is in South America an extensive family of these insects, the Heliconidæ, which are in many respects very remarkable. They are so abundant and characteristic in all the woody portions of the American tropics, that in almost every locality they will be seen more frequently than any other butterflies. They are distinguished by very elongate wings, body, and antennæ, and are exceedingly beautiful and varied in their colours; spots and patches of yellow, red, or pure white upon a black, blue, or brown ground, being most general. They frequent the forests chiefly, and all fly slowly and weakly; yet although they are so conspicuous, and could certainly be caught by insectivorous birds more easily than almost any other insects, their great abundance all over the wide region they inhabit shows that they are not so persecuted. It is to be especially remarked also, that they possess no adaptive colouring to protect them during repose, for the under side of their wings presents the same, or at least an equally conspicuous colouring as the upper side; and they may be observed after sunset suspended at the end of twigs and leaves where they have taken up their station for the night, fully exposed to the attacks of enemies if they have any. These beautiful insects possess, however, a strong pungent semi-aromatic or medicinal odour, which seems to pervade all the juices of their system. When the entomologist squeezes the breast of one of them between his fingers to kill it, a yellow liquid exudes which stains the skin, and the smell of which can only be got rid of by time and repeated washings. Here we have probably the cause of their immunity from attack, since there is a great deal of evidence to show that certain insects are so disgusting to birds that they will under no circumstances touch them. Mr. Stainton has observed that a brood of young turkeys greedily devoured all the worthless moths he had amassed in a night's "sugaring," yet one after another seized and rejected a single white moth which happened to be among them. Young pheasants and partridges which eat many kinds of caterpillars seem to have an absolute dread of that of the common currant moth, which they will never touch, and tomtits as well as other small birds appear never to eat the same species. In the case of the Heliconidæ, however, we have some direct evidence to the same effect. In the Brazilian forests there are great numbers of insectivorous birds — as jacamars, trogons, and puffbirds — which catch insects on the wing, and that they destroy many butterflies is indicated by the fact that the wings of these insects are often found on the ground where their bodies have been devoured. But among these there are no wings of Heliconidæ, while those of the large showy Nymphalidæ, which have a much swifter flight, are often met with. Again, a gentleman who had recently returned from Brazil stated at a meeting of the Entomological Society that he once observed a pair of puffbirds catching butterflies, which they brought to their nest to feed their young; yet during half an hour they never brought one of the Heliconidæ, which were flying lazily about in great numbers, and which they could have captured more easily than any others. It was this circumstance that led Mr. Belt to observe them so long, as he could not understand why the most common insects should be altogether passed by. Mr. Bates also tells us that he never saw them molested by lizards or predacious flies, which often pounce on other butterflies.

If, therefore, we accept it as highly probable (if not proved) that the Heliconidæ are very greatly protected from attack by their peculiar odour and taste, we find it much more easy to understand their chief characteristics — their great abundance, their slow flight, their gaudy colours, and the entire absence of protective tints on their under surfaces. This property places them somewhat in the position of those curious wingless birds of oceanic islands, the dodo, the apteryx, and the moas, which are with great reason supposed to have lost the power of flight on account of the absence of carnivorous quadrupeds. Our butterflies have been protected in a different way, but quite as effectually; and the result has been that as there has been nothing to escape from, there has been no weeding out of slow flyers, and as there has been nothing to hide from, there has been no extermination of the bright-coloured varieties, and no preservation of such as tended to assimilate with surrounding objects.

Now let us consider how this kind of protection must act. Tropical insectivorous birds very frequently sit on dead branches of a lofty tree, or on those which overhang forest paths, gazing intently around, and darting off at intervals to seize an insect at a considerable distance, which they generally return to their station to devour. If a bird began by capturing the slow-flying, conspicuous Heliconidæ, and found them always so disagreeable that it could not eat them, it would after a very few trials leave off catching them at all; and their whole appearance, form, colouring, and mode of flight is so peculiar, that there can be little doubt birds would soon learn to distinguish them at a long distance, and never waste any time in pursuit of them. Under these circumstances, it is evident that any other butterfly of a group which birds were accustomed to devour, would be almost equally well protected by closely resembling a Heliconia externally, as if it acquired also the disagreeable odour; always supposing that there were only a few of them among a great number of the Heliconias. If the birds could not distinguish the two kinds externally, and there were on the average only one eatable among fifty uneatable, they would soon give up seeking for the eatable ones, even if they knew them to exist. If, on the other hand, any particular butterfly of an eatable group acquired the disagreeable taste of the Heliconias while it retained the characteristic form and colouring of its own group, this would be really of no use to it whatever; for the birds would go on catching it among its eatable allies (compared with which it would rarely occur), it would be wounded and disabled, even if rejected, and its increase would thus be as effectually checked as if it were devoured. It is important, therefore, to understand that if any one genus of an extensive family of eatable butterflies were in danger of extermination from insect-eating birds, and if two kinds of variation were going on among them, some individuals possessing a slightly disagreeable taste, others a slight resemblance to the Heliconidæ, this latter quality would be much more valuable than the former. The change in flavour would not at all prevent the variety from being captured as before, and it would almost certainly be thoroughly disabled before being rejected. The approach in colour and form to the Heliconidæ, however, would be at the very first a positive, though perhaps a slight advantage; for although at short distances this variety would be easily distinguished and devoured, yet at a longer distance it might be mistaken for one of the uneatable group, and so be passed by and gain another day's life, which might in many cases be sufficient for it to lay a quantity of eggs and leave a numerous progeny, many of which would inherit the peculiarity which had been the safeguard of their parent.

Now, this hypothetical case is exactly realized in South America. Among the white butterflies forming the family Pieridæ (many of which do not greatly differ in appearance from our own cabbage butterflies) is a genus of rather small size (Leptalis), some species of which are white like their allies, while the larger number exactly resemble the Heliconidæ in the form and colouring of the wings. It must always be remembered that these two families are as absolutely distinguished from each other by structural characters as are the carnivora and the ruminants among quadrupeds, and that an entomologist can always distinguish the one from the other by the structure of the feet, just as certainly as a zoologist can tell a bear from a buffalo by the skull or by a tooth. Yet the resemblance of a species of the one family to another species in the other family was often so great, that both Mr. Bates and myself were many times deceived at the time of capture, and did not discover the distinctness of the two insects till a closer examination detected their essential differences. During his residence of eleven years in the Amazon valley, Mr. Bates found a number of species or varieties of Leptalis, each of which was a more or less exact copy of one of the Heliconidæ of the district it inhabited; and the results of his observations are embodied in a paper published in the Linnean Transactions, in which he first explained the phenomena of "mimicry" as the result of natural selection, and showed its identity in cause and purpose with protective resemblance to vegetable or inorganic forms.

The imitation of the Heliconidæ by the Leptalides is carried out to a wonderful degree in form as well as in colouring. The wings have become elongated to the same extent, and the antennæ and abdomen have both become lengthened, to correspond with the unusual condition in which they exist in the former family. In colouration there are several types in the different genera of Heliconidæ. The genus Mechanitis is generally of a rich semi-transparent brown, banded with black and yellow; Methona is of large size, the wings transparent like horn, and with black transverse bands; while the delicate Ithomias are all more or less transparent, with black veins and borders, and often with marginal and transverse bands of orange red. These different forms are all copied by the various species of Leptalis, every band and spot and tint of colour, and the various degrees of transparency, being exactly reproduced. As if to derive all the benefit possible from this protective mimicry, the habits have become so modified that the Leptalides generally frequent the very same spots as their models, and have the same mode of flight; and as they are always very scarce (Mr. Bates estimating their numbers at about one to a thousand of the group they resemble), there is hardly a possibility of their being found out by their enemies. It is also very remarkable that in almost every case the particular Ithomias and other species of Heliconidæ which they resemble, are noted as being very common species, swarming in individuals, and found over a wide range of country. This indicates antiquity and permanence in the species, and is exactly the condition most essential both to aid in the development of the resemblance, and to increase its utility.

But the Leptalides are not the only insects who have prolonged their existence by imitating the great protected group of Heliconidæ;— a genus of quite another family of most lovely small American butterflies, the Erycinidæ, and three genera of diurnal moths, also present species which often mimic the same dominant forms, so that some, as Ithomia ilerdina of St. Paulo, for instance, have flying with them a few individuals of three widely different insects, which are yet disguised with exactly the same form, colour, and markings, so as to be quite undistinguishable when upon the wing. Again, the Heliconidæ are not the only group that are imitated, although they are the most frequent models. The black and red group of South American Papilios, and the handsome Erycinian genus Stalachtis, have also a few who copy them; but this fact offers no difficulty, since these two groups are almost as dominant as the Heliconidæ. They both fly very slowly, they are both conspicuously coloured, and they both abound in individuals; so that there is every reason to believe that they possess a protection of a similar kind to the Heliconidæ, and that it is therefore equally an advantage to other insects to be mistaken for them. There is also another extraordinary fact that we are not yet in a position clearly to comprehend: some groups of the Heliconidæ themselves mimic other groups. Species of Heliconia mimic Mechanitis, and every species of Napeogenes mimics some other Heliconideous butterfly. This would seem to indicate that the distasteful secretion is not produced alike by all members of the family, and that where it is deficient protective imitation comes into play. It is this, perhaps, that has caused such a general resemblance among the Heliconidæ, such a uniformity of type with great diversity of colouring, since any aberration causing an insect to cease to look like one of the family would inevitably lead to its being attacked, wounded, and exterminated, even although it was not eatable.

In other parts of the world an exactly parallel series of facts have been observed. The Danaidæ and the Acræidæ of the Old World tropics form in fact one great group with the Heliconidæ. They have the same general form, structure, and habits: they possess the same protective odour, and are equally abundant in individuals, although not so varied in colour, blue and white spots on a black ground being the most general pattern. The insects which mimic these are chiefly Papilios, and Diadema, a genus allied to our peacock and tortoiseshell butterflies. In tropical Africa there is a peculiar group of the genus Danais, characterized by dark-brown and bluish-white colours, arranged in bands or stripes. One of these, Danais niavius, is exactly imitated both by Papilio hippocoon and by Diadema anthedon; another, Danais echeria, by Papilio cenea; and in Natal a variety of the Danais is found having a white spot at the tip of wings, accompanied by a variety of the Papilio bearing a corresponding white spot. Acræa gea is copied in its very peculiar style of colouration by the female of Papilio cynorta, by Panopæa hirce, and by the female of Elymnias phegea. Acræa euryta of Calabar has a female variety of Panopea hirce from the same place which exactly copies it; and Mr. Trimen, in his paper on Mimetic Analogies among African Butterflies, published in the Transactions of the Linnæan Society for 1868, gives a list of no less than sixteen species and varieties of Diadema and its allies, and ten of Papilio, which in their colour and markings are perfect mimics of species or varieties of Danais or Acræa which inhabit the same districts.

Passing on to India, we have Danais tytia, a butterfly with semi-transparent bluish wings and a border of rich reddish brown. This remarkable style of colouring is exactly reproduced in Papilio agestor and in Diadema nama, and all three insects not unfrequently come together in collections made at Darjeeling. In the Philippine Islands the large and curious Idea leuconöe with its semi-transparent white wings, veined and spotted with black, is copied by the rare Papilio idæoides from the same islands.

In the Malay archipelago the very common and beautiful Euploea midamus is so exactly mimicked by two rare Papilios (P. paradoxa and P. ænigma) that I generally caught them under the impression that they were the more common species; and the equally common and even more beautiful Euploea rhadamanthus, with its pure white bands and spots on a ground of glossy blue and black, is reproduced in the Papilio caunus. Here also there are species of Diadema imitating the same group in two or three instances; but we shall have to adduce these further on in connexion with another branch of the subject.

It has been already mentioned that in South America there is a group of Papilios which have all the characteristics of a protected race, and whose peculiar colours and markings are imitated by other butterflies not so protected. There is just such a group also in the East, having very similar colours and the same habits, and these also are mimicked by other species in the same genus not closely allied to them, and also by a few of other families. Papilio hector, a common Indian butterfly of a rich black colour spotted with crimson, is so closely copied by Papilio romulus, that the latter insect has been thought to be its female. A close examination shows, however, that it is essentially different, and belongs to another section of the genus. Papilio antiphus and P. diphilus, black swallow-tailed butterflies with cream-coloured spots, are so well imitated by varieties of P. theseus, that several writers have classed them as the same species. Papilio liris, found only in the island of Timor, is accompanied there by P. ænomaus, the female of which so exactly resembles it that they can hardly be separated in the cabinet, and on the wing are quite undistinguishable. But one of the most curious cases is the fine yellow-spotted Papilio cöon, which is unmistakeably imitated by the female tailed form of Papilio memnon. These are both from Sumatra; but in North India P. cöon is replaced by another species, which has been named P. doubledayi, having red spots instead of yellow; and in the same district the corresponding female tailed form of Papilio androgeus, sometimes considered a variety of P. memnon, is similarly red-spotted. Mr. Westwood has described some curious day-flying moths (Epicopeia) from North India, which have the form and colour of Papilios of this section, and two of these are very good imitations of Papilio polydorus and Papilio varuna, also from North India.

Almost all these cases of mimicry are from the tropics, where the forms of life are more abundant, and where insect development especially is of unchecked luxuriance; but there are also one or two instances in temperate regions. In North America, the large and handsome red and black butterfly Danais erippus is very common; and the same country is inhabited by Limenitis archippus, which closely resembles the Danais, while it differs entirely from every species of its own genus.

The only case of probable mimicry in our own country is the following:— A very common white moth (Spilosoma menthastri) was found by Mr. Stainton to be rejected by young turkeys among hundreds of other moths on which they greedily fed. Each bird in succession took hold of this moth and threw it down again, as if too nasty to eat. Mr. Jenner Weir also found that this moth was refused by the Bullfinch, Chaffinch, Yellow Hammer, and Red Bunting, but eaten after much hesitation by the Robin. We may therefore fairly conclude that this species would be disagreeable to many other birds, and would thus have an immunity from attack, which may be the cause of its great abundance and of its conspicuous white colour. Now it is a curious thing that there is another moth, Diaphora mendica, which appears about the same time, and whose female only is white. It is about the same size as Spilosoma menthastri, and sufficiently resembles it in the dusk, and this moth is much less common. It seems very probable, therefore, that these species stand in the same relation to each other as the mimicking butterflies of various families do to the Heliconidæ and Danaidæ. It would be very interesting to experiment on all white moths, to ascertain if those which are most common are generally rejected by birds. It may be anticipated that they would be so, because white is the most conspicuous of all colours for nocturnal insects, and had they not some other protection would certainly be very injurious to them.

Lepidoptera mimicking other Insects.

In the preceding cases we have found Lepidoptera imitating other species of the same order, and such species only as we have good reason to believe were free from the attacks of many insectivorous creatures; but there are other instances in which they altogether lose the external appearance of the order to which they belong, and take on the dress of bees or wasps — insects which have an undeniable protection in their stings. The Sesiidæ and Ægeriidæ, two families of day-flying moths, are particularly remarkable in this respect, and a mere inspection of the names given to the various species shows how the resemblance has struck everyone. We have apiformis, vespiforme, ichneumoniforme, scoliæforme, sphegiforme (bee-like, wasp-like, ichneumon-like, &c.) and many others, all indicating a resemblance to stinging Hymenoptera. In Britain we may particularly notice Sesia bombiliformis, which very closely resembles the male of the large and common humble bee, Bombus hortorum; Sphecia craboniforme, which is coloured like a hornet, and is (on the authority of Mr. Jenner Weir) much more like it when alive than when in the cabinet, from the way in which it carries its wings; and the currant clear-wing, Trochilium tipuliforme, which resembles a small black wasp (Odynerus sinuatus) very abundant in gardens at the same season. It has been so much the practice to look upon these resemblances as mere curious analogies playing no part in the economy of nature, that we have scarcely any observations of the habits and appearance when alive of the hundreds of species of these groups in various parts of the world, or how far they are accompanied by Hymenoptera, which they specifically resemble. There are many species in India (like those figured by Professor Westwood in his "Oriental Entomology") which have the hind legs very broad and densely hairy, so as exactly to imitate the brush-legged bees (Scopulipedes) which abound in the same country. In this case we have more than mere resemblance of colour, for that which is an important functional structure in the one group is imitated in another whose habits render it perfectly useless.

Mimicry among Beetles.

It may fairly be expected that if these imitations of one creature by another really serve as a protection to weak and decaying species, instances of the same kind will be found among other groups than the Lepidoptera; and such is the case, although they are seldom so prominent and so easily recognised as those already pointed out as occurring in that order. A few very interesting examples may, however, be pointed out in most of the other orders of insects. The Coleoptera or beetles that imitate other Coleoptera of distinct groups are very numerous in tropical countries, and they generally follow the laws already laid down as regulating these phenomena. The insects which others imitate always have a special protection, which leads them to be avoided as dangerous or uneatable by small insectivorous animals; some have a disgusting taste (analogous to that of the Heliconidæ); others have such a hard and stony covering that they cannot be crushed or digested; while a third set are very active, and armed with powerful jaws, as well as having some disagreeable secretion. Some species of Eumorphidæ and Hispidæ, small flat or hemispherical beetles which are exceedingly abundant, and have a disagreeable secretion, are imitated by others of the very distinct group of Longicornes (of which our common musk-beetle may be taken as an example). The extraordinary little Cyclopeplus batesii, belongs to the same sub-family of this group as the Onychocerus scorpio and O. concentricus, which have already been adduced as imitating with such wonderful accuracy the bark of the trees they habitually frequent; but it differs totally in outward appearance from every one of its allies, having taken upon itself the exact shape and colouring of a globular Corynomalus, a little stinking beetle with clubbed antennæ. It is curious to see how these clubbed antennæ are imitated by an insect belonging to a group with long slender antennæ. The sub-family Anisocerinæ, to which Cyclopeplus belongs, is characterised by all its members possessing a little knob or dilatation about the middle of the antennæ. This knob is considerably enlarged in C. batesii, and the terminal portion of the antennæ beyond it is so small and slender as to be scarcely visible, and thus an excellent substitute is obtained for the short clubbed antennæ of the Corynomalus. Erythroplatis corallifer is another curious broad flat beetle, that no one would take for a Longicorn, since it almost exactly resembles Cephalodonta spinipes, one of the commonest of the South American Hispidæ; and what is still more remarkable, another Longicorn of a distinct group, Streptolabis hispoides, was found by Mr. Bates, which resembles the same insect with equal minuteness,— a case exactly parallel to that among butterflies, where species of two or three distinct groups mimicked the same Heliconia. Many of the soft-winged beetles (Malacoderms) are excessively abundant in individuals, and it is probable that they have some similar protection, more especially as other species often strikingly resemble them. A Longicorn beetle, Pæciloderma terminale, found in Jamaica, is coloured exactly in the same way as a Lycus (one of the Malacoderms) from the same island. Eroschema poweri, a Longicorn from Australia, might certainly be taken for one of the same group, and several species from the Malay Islands are equally deceptive. In the Island of Celebes I found one of this group, having the whole body and elytra of a rich deep blue colour, with the head only orange; and in company with it an insect of a totally different family (Eucnemidæ) with identically the same colouration, and of so nearly the same size and form as to completely puzzle the collector on every fresh occasion of capturing them. I have been recently informed by Mr. Jenner Weir, who keeps a variety of small birds, that none of them will touch our common "soldiers and sailors" (species of Malacoderms), thus confirming my belief that they were a protected group, founded on the fact of their being at once very abundant, of conspicuous colours, and the objects of mimicry.

There are a number of the larger tropical weevils which have the elytra and the whole covering of the body so hard as to be a great annoyance to the entomologist, because in attempting to transfix them the points of his pins are constantly turned. I have found it necessary in these cases to drill a hole very carefully with the point of a sharp penknife before attempting to insert a pin. Many of the fine long-antennæd Anthribidæ (an allied group) have to be treated in the same way. We can easily understand that after small birds have in vain attempted to eat these insects, they should get to know them by sight, and ever after leave them alone, and it will then be an advantage for other insects which are comparatively soft and eatable, to be mistaken for them. We need not be surprised, therefore, to find that there are many Longicorns which strikingly resemble the "hard beetles" of their own district. In South Brazil, Acanthotritus dorsalis is strikingly like a Curculio of the hard genus Heiliplus, and Mr. Bates assures me that he found Gymnocerus cratosomoides (a Longicorn) on the same tree with a hard Cratosomus (a weevil), which it exactly mimics. Again, the pretty Longicorn, Phacellocera batesii, mimics one of the hard Anthribidæ of the genus Ptychoderes, having long slender antennæ. In the Moluccas we find Cacia anthriboides, a small Longicorn which might be easily mistaken for a very common species of Anthribidæ found in the same districts; and the very rare Capnolymma stygium closely imitates the common Mecocerus gazella, which abounded where it was taken. Doliops curculionoides and other allied Longicorns from the Philippine Islands most curiously resemble, both in form and colouring, the brilliant Pachyrhynchi,— Curculionidæ, which are almost peculiar to that group of islands. The remaining family of Coleoptera most frequently imitated is the Cicindelidæ. The rare and curious Longicorn, Collyrodes lacordairei, has exactly the form and colouring of the genus Collyris, while an undescribed species of Heteromera is exactly like a Therates, and was taken running on the trunks of trees, as is the habit of that group. There is one curious example of a Longicorn mimicking a Longicorn, like the Papilios and Heliconidæ which mimic their own allies. Agnia fasciata, belonging to the sub-family Hypselominæ, and Nemophas grayi, belonging to the Lamiinæ, were taken in Amboyna on the same fallen tree at the same time, and were supposed to be the same species till they were more carefully examined, and found to be structurally quite different. The colouring of these insects is very remarkable, being rich steel-blue black, crossed by broad hairy bands of orange buff, and out of the many thousands of known species of Longicorns they are probably the only two which are so coloured. The

Nemophas grayi is the larger, stronger, and better armed insect, and belongs to a more widely spread and dominant group, very rich in species and individuals, and is therefore most probably the subject of mimicry by the other species.

Beetles mimicking other Insects.

We will now adduce a few cases in which beetles imitate other insects, and insects of other orders imitate beetles.

Charis melipona, a South American Longicorn of the family Necydalidæ, has been so named from its resemblance to a small bee of the genus Melipona. It is one of the most remarkable cases of mimicry, since the beetle has the thorax and body densely hairy like the bee, and the legs are tufted in a manner most unusual in the order Coleoptera. Another Longicorn, Odontocera odyneroides, has the abdomen banded with yellow, and constricted at the base, and is altogether so exactly like a small common wasp of the genus Odynerus, that Mr. Bates informs us he was afraid to take it out of his net with his fingers for fear of being stung. Had Mr. Bates's taste for insects been less omnivorous than it was, the beetle's disguise might have saved it from his pin, as it had no doubt often done from the beak of hungry birds. A larger insect, Sphecomorpha chalybea, is exactly like one of the large metallic blue wasps, and like them has the abdomen connected with the thorax by a pedicel, rendering the deception most complete and striking. Many Eastern species of Longicorns of the genus Oberea, when on the wing exactly resemble Tenthredinidæ, and many of the small species of Hesthesis run about on timber, and cannot be distinguished from ants. There is one genus of South American Longicorns that appears to mimic the shielded bugs of the genus Scutellera. The Gymnocerous capucinus is one of these, and is very like Pachyotris fabricii, one of the Scutelleridæ. The beautiful Gymnocerous dulcissimus is also very like the same group of insects, though there is no known species that exactly corresponds to it; but this is not to be wondered at, as the tropical Hemiptera have been comparatively so little cared for by collectors.

Insects mimicking Species of other Orders.

The most remarkable case of an insect of another order mimicking a beetle is that of the Condylodera tricondyloides, one of the cricket family from the Philippine Islands, which is so exactly like a Tricondyla (one of the tiger beetles), that such an experienced entomologist as Professor Westwood placed it among them in his cabinet, and retained it there a long time before he discovered his mistake! Both insects run along the trunks of trees, and whereas Tricondylas are very plentiful, the insect that mimics it is, as in all other cases, very rare. Mr. Bates also informs us that he found at Santarem on the Amazon, a species of locust which mimicked one of the tiger beetles of the genus Odontocheila, and was found on the same trees which they frequented.

There are a considerable number of Diptera, or two-winged flies, that closely resemble wasps and bees, and no doubt derive much benefit from the wholesome dread which those insects excite. The Midas dives, and other species of large Brazilian flies, have dark wings and metallic blue elongate bodies, resembling the large stinging Sphegidæ of the same country; and a very large fly of the genus Asilus has black-banded wings and the abdomen tipped with rich orange, so as exactly to resemble the fine bee Euglossa dimidiata, and both are found in the same parts of South America. We have also in our own country species of Bombylius which are almost exactly like bees. In these cases the end gained by the mimicry is no doubt freedom from attack, but it has sometimes an altogether different purpose. There are a number of parasitic flies whose larvæ feed upon the larvæ of bees, such as the British genus Volucella and many of the tropical Bombylii, and most of these are exactly like the particular species of bee they prey upon, so that they can enter their nests unsuspected to deposit their eggs. There are also bees that mimic bees. The cuckoo bees of the genus Nomada are parasitic on the Andrenidæ, and they resemble either wasps or species of Andrena; and the parasitic humble-bees of the genus Apathus almost exactly resemble the species of humble-bees in whose nests they are reared. Mr. Bates informs us that he found numbers of these "cuckoo" bees and flies on the Amazon, which all wore the livery of working bees peculiar to the same country.

There is a genus of small spiders in the tropics which feed on ants, and they are exactly like ants themselves, which no doubt gives them more opportunity of seizing their prey; and Mr. Bates found on the Amazon a species of Mantis which exactly resembled the white ants which it fed upon, as well as several species of crickets (Scaphura), which resembled in a wonderful manner different sand-wasps of large size, which are constantly on the search for crickets with which to provision their nests.

Perhaps the most wonderful case of all is the large caterpillar mentioned by Mr. Bates, which startled him by its close resemblance to a small snake. The first three segments behind the head were dilatable at the will of the insect, and had on each side a large black pupillated spot, which resembled the eye of the reptile. Moreover, it resembled a poisonous viper, not a harmless species of snake, as was proved by the imitation of keeled scales on the crown produced by the recumbent feet, as the caterpillar threw itself backward!

The attitudes of many of the tropical spiders are most

extraordinary and deceptive, but little attention has been paid to them. They often mimic other insects, and some, Mr. Bates assures us, are exactly like flower buds, and take their station in the axils of leaves, where they remain motionless waiting for their prey.

Cases of Mimicry among the Vertebrata.

Having thus shown how varied and extraordinary are the modes in which mimicry occurs among insects, we have now to enquire if anything of the same kind is to be observed among vertebrated animals. When we consider all the conditions necessary to produce a good deceptive imitation, we shall see at once that such can very rarely occur in the higher animals, since they possess none of those facilities for the almost infinite modifications of external form which exist in the very nature of insect organization. The outer covering of insects being more or less solid and horny, they are capable of almost any amount of change of form and appearance without any essential modification internally. In many groups the wings give much of the character, and these organs may be much modified both in form and colour without interfering with their special functions. Again, the number of species of insects is so great, and there is such diversity of form and proportion in every group, that the chances of an accidental approximation in size, form, and colour, of one insect to another of a different group, are very considerable; and it is these chance approximations that furnish the basis of mimicry, to be continually advanced and perfected by the survival of those varieties only which tend in the right direction.

In the Vertebrata, on the contrary, the skeleton being internal the external form depends almost entirely on the proportions and arrangement of that skeleton, which again is strictly adapted to the functions necessary for the well-being of the animal. The form cannot therefore be rapidly modified by variation, and the thin and flexible integument will not admit of the development of such strange protuberances as occur continually in insects. The number of species of each group in the same country is also comparatively small, and thus the chances of that first accidental resemblance which is necessary for natural selection to work upon are much diminished. We can hardly see the possibility of a mimicry by which the elk could escape from the wolf, or the buffalo from the tiger. There is, however, in one group of Vertebrata such a general similarity of form, that a very slight modification, if accompanied by identity of colour, would produce the necessary amount of resemblance; and at the same time there exist a number of species which it would be advantageous for others to resemble, since they are armed with the most fatal weapons of offence. We accordingly find that reptiles furnish us with a very remarkable and instructive case of true mimicry.

Mimicry among Snakes.

There are in tropical America a number of venomous snakes of the genus Elaps, which are ornamented with brilliant colours disposed in a peculiar manner. The ground colour is generally bright red, on which are black bands of various widths and sometimes divided into two or three by yellow rings. Now, in the same country are found several genera of harmless snakes, having no affinity whatever with the above, but coloured exactly the same. For example, the poisonous Elaps fulvius often occurs in Guatemala with simple black bands on a coral-red ground; and in the same country is found the harmless snake Pliocerus equalis, coloured and banded in identically the same manner. A variety of Elaps corallinus has the black bands narrowly bordered with yellow on the same red ground colour, and a harmless snake, Homalocranium semicinctum, has exactly the same markings, and both are found in Mexico. The deadly Elaps lemniscatus has the black bands very broad, and each of them divided into three by narrow yellow rings; and this again is exactly copied by a harmless snake, Pliocerus elapoides, which is found along with its model in Mexico.

But, more remarkable still, there is in South America a third group of snakes, the genus Oxyrhopus, doubtfully venomous, and having no immediate affinity with either of the preceding, which has also the same curious distribution of colours, namely, variously disposed rings of red, yellow, and black; and there are some cases in which species of all three of these groups similarly marked inhabit the same district. For example, Elaps mipartitus has single black rings very close together. It inhabits the west side of the Andes, and in the same districts occur Pliocerus euryzonus and Oxyrhopus petolarius, which exactly copy its pattern. In Brazil Elaps lemniscatus is copied by Oxyrhopus trigeminus, both having black rings disposed in threes. In Elaps hemiprichii the ground colour appears to be black, with alternations of two narrow yellow bands and a broader red one; and of this pattern again we have an exact double in Oxyrhopus formosus, both being found in many localities of tropical South America.

What adds much to the extraordinary character of these resemblances is the fact, that nowhere in the world but in America are there any snakes at all which have this style of colouring. Dr. Gunther, of the British Museum, who has kindly furnished some of the details here referred to, assures me that this is the case; and that red, black, and yellow rings occur together on no other snakes in the world but on Elaps and the species which so closely resemble it. In all these cases, the size and form as well as the colouration, are so much alike, that none but a naturalist would distinguish the harmless from the poisonous species.

Many of the small tree-frogs are no doubt also mimickers. When seen in their natural attitudes, I have been often unable to distinguish them from beetles or other insects sitting upon leaves, but regret to say I neglected to observe what species or groups they most resembled, and the subject does not yet seem to have attracted the attention of naturalists abroad.

Mimicry among Birds.

In the class of birds there are a number of cases that make some approach to mimicry, such as the resemblance of the cuckoos, a weak and defenceless group of birds, to hawks and Gallinaceæ. There is, however, one example which goes much further than this, and seems to be of exactly the same nature as the many cases of insect mimicry which have been already given. In Australia and the Moluccas there is a genus of honeysuckers called Tropidorhynchus, good sized birds, very strong and active, having powerful grasping claws and long, curved, sharp beaks. They assemble together in groups and small flocks, and they have a very loud bawling note, which can be heard at a great distance, and serves to collect a number together in time of danger. They are very plentiful and very pugnacious, frequently driving away crows, and even hawks, which perch on a tree where a few of them are assembled. They are all of rather dull and obscure colours. Now in the same countries there is a group of orioles, forming the genus Mimeta, much weaker birds, which have lost the gay colouring of their allies the golden orioles, being usually olive-green or brown; and in several cases these most curiously resemble the Tropidorhynchus of the same island. For example, in the island of Bouru is found the Tropidorhynchus bouruensis, of a dull earthy colour, and the Mimeta bouruensis, which resembles it in the following particulars:— The upper and under surfaces of the two birds are exactly of the same tints of dark and light brown; the Tropidorhynchus has a large bare black patch round the eyes; this is copied in the Mimeta by a patch of black feathers. The top of the head of the Tropidorhynchus has a scaly appearance from the narrow scale-formed feathers, which are imitated by the broader feathers of the Mimeta having a dusky line down each. The Tropidorhynchus has a pale ruff formed of curious recurved feathers on the nape (which has given the whole genus the name of Friar birds); this is represented in the Mimeta by a pale band in the same position. Lastly, the bill of the Tropidorhynchus is raised into a protuberant keel at the base, and the Mimeta has the same character, although it is not a common one in the genus. The result is, that on a superficial examination the birds are identical, although they have important structural differences, and cannot be placed near each other in any natural arrangement. As a proof that the resemblance is really deceptive, it may be mentioned that the Mimeta is figured and described as a honeysucker in the costly "Voyage de l'Astrolabe," under the name of Philedon bouruensis!

Passing to the island of Ceram, we find allied species of both genera. The Tropidorhynchus subcornutus is of an earthy brown colour washed with yellow ochre, with bare orbits, dusky cheeks, and the usual pale recurved nape-ruff. The Mimeta forsteni is absolutely identical in the tints of every part of the body, the details of which are imitated in the same manner as in the Bouru birds already described. In two other islands there is an approximation towards mimicry, although it is not so perfect as in the two preceding cases. In Timor the Tropidorhynchus timoriensis is of the usual earthy brown above, with the nape-ruff very prominent, the cheeks black, the throat nearly white, and the whole under surface pale whitish brown. These various tints are all well reproduced in Mimeta virescens, the chief want of exact imitation being that the throat and breast of the Tropidorhynchus has a very scaly appearance, being covered with rigid pointed feathers which are not imitated in the Mimeta, although there are signs of faint dusky spots which may easily furnish the groundwork of a more exact imitation by the continued survival of favourable variations in the same direction. There is also a large knob at the base of the bill of the Tropidorhynchus which is not at all imitated by the Mimeta. In the island of Morty (north of Gilolo) there exists the Tropidorhynchus fuscicapillus, of a dark sooty brown colour, especially on the head, while the under parts are rather lighter, and the characteristic ruff of the nape is wanting. Now it is curious that in the adjacent island of Gilolo should be found the Mimeta phæochromus, the upper surface of which is of exactly the same dark sooty tint as the Tropidorhynchus, and is the only known species that is of such a dark colour. The under side is not quite light enough, but it is a good approximation. This Mimeta is a rare bird, and may very probably exist in Morty, though not yet found there; or, on the other hand, recent changes in physical geography may have led to the restriction of the Tropidorhynchus to that island, where it is very common.

Here, then, we have two cases of perfect mimicry and two others of good approximation, occurring between species of the same two genera of birds; and in three of these cases the pairs that resemble each other are found together in the same island, and to which they are peculiar. In all these cases the Tropidorhynchus is rather larger than the Mimeta, but the difference is not beyond the limits of variation in species, and the two genera are somewhat alike in form and proportion. There are, no doubt, some special enemies by

73

which many small birds are attacked, but which are afraid of the Tropidorhynchus (probably some of the hawks), and thus it becomes advantageous for the weak Mimeta to resemble the strong, pugnacious, noisy, and very abundant Tropidorhynchus.

My friend, Mr. Osbert Salvin, has given me another interesting case of bird mimicry. In the neighbourhood of Rio Janeiro is found an insect-eating hawk (Harpagus diodon), and in the same district a bird-eating hawk (Accipiter pileatus) which closely resembles it. Both are of the same ashy tint beneath, with the thighs and under wing-coverts reddish brown, so that when on the wing and seen from below they are undistinguishable. The curious point, however, is that the Accipiter has a much wider range than the Harpagus, and in the regions where the insect-eating species is not found it no longer resembles it, the under wing-coverts varying to white; thus indicating that the red-brown colour is kept true by its being useful to the Accipiter to be mistaken for the insect-eating species, which birds have learnt not to be afraid of.

Mimicry among Mammals.

Among the Mammalia the only case which may be true mimicry is that of the insectivorous genus Cladobates, found in the Malay countries, several species of which very closely resemble squirrels. The size is about the same, the long bushy tail is carried in the same way, and the colours are very similar. In this case the use of the resemblance must be to enable the Cladobates to approach the insects or small birds on which it feeds, under the disguise of the harmless fruit-eating squirrel.

Objections to Mr. Bates' Theory of Mimicry.

Having now completed our survey of the most prominent and remarkable cases of mimicry that have yet been noticed, we must say something of the objections that have been made to the theory of their production given by Mr. Bates, and which we have endeavoured to illustrate and enforce in the preceding pages. Three counter explanations have been proposed. Professor Westwood admits the fact of the mimicry and its probable use to the insect, but maintains that each species was created a mimic for the purpose of the protection thus afforded it. Mr. Andrew Murray, in his paper on the "Disguises of Nature," inclines to the opinion that similar conditions of food and of surrounding circumstances have acted in some unknown way to produce the resemblances; and when the subject was discussed before the Entomological Society of London, a third objection was added — that heredity or the reversion to ancestral types of form and colouration, might have produced many of the cases of mimicry.

Against the special creation of mimicking species there are all the objections and difficulties in the way of special creation in other cases, with the addition of a few that are peculiar to it. The most obvious is, that we have gradations of mimicry and of protective resemblance — a fact which is strongly suggestive of a natural process having been at work. Another very serious objection is, that as mimicry has been shown to be useful only to those species and groups which are rare and probably dying out, and would cease to have any effect should the proportionate abundance of the two species be reversed, it follows that on the special-creation theory the one species must have been created plentiful, the other rare; and, notwithstanding the many causes that continually tend to alter the proportions of species, these two species must have always been specially maintained at their respective proportions, or the very purpose for which they each received their peculiar characteristics would have completely failed. A third difficulty is, that although it is very easy to understand how mimicry may be brought about by variation and the survival of the fittest, it seems a very strange thing for a Creator to protect an animal by making it imitate another, when the very assumption of a Creator implies his power to create it so as to require no such circuitous protection. These appear to be fatal objections to the application of the special-creation theory to this particular case.

The other two supposed explanations, which may be shortly expressed as the theories of "similar conditions" and of "heredity," agree in making mimicry, where it exists, an adventitious circumstance not necessarily connected with the well-being of the mimicking species. But several of the most striking and most constant facts which have been adduced, directly contradict both those hypotheses. The law that mimicry is confined to a few groups only is one of these, for "similar conditions" must act more or less on all groups in a limited region, and "heredity" must influence all groups related to each other in an equal degree. Again, the general fact that those species which mimic others are rare, while those which are imitated are abundant, is in no way explained by either of these theories, any more than is the frequent occurrence of some palpable mode of protection in the imitated species. "Reversion to an ancestral type" no way explains why the imitator and the imitated always inhabit the very same district, whereas allied forms of every degree of nearness and remoteness generally inhabit different countries, and often different quarters of the globe; and neither it, nor "similar conditions," will account for the likeness between species of distinct groups being superficial only — a disguise, not a true resemblance; for the imitation of bark, of leaves, of sticks, of dung; for the resemblance between species in different orders, and even different classes and sub-kingdoms; and finally, for the graduated series of the phenomena, beginning with a general harmony and adaptation of tint in autumn and winter moths and in arctic and desert animals, and ending with those complete cases of detailed mimicry which not only deceive predacious animals, but puzzle the most experienced insect collectors and the most learned entomologists.

Mimicry by Female Insects only.

But there is yet another series of phenomena connected with this subject, which considerably strengthens the view here adopted, while it seems quite incompatible with either of the other hypotheses; namely, the relation of protective colouring and mimicry to the sexual differences of animals. It will be clear to every one that if two animals, which as regards "external conditions" and "hereditary descent," are exactly alike, yet differ remarkably in colouration, one resembling a protected species and the other not, the resemblance that exists in one only can hardly be imputed to the influence of external conditions or as the effect of heredity. And if, further, it can be proved that the one requires protection more than the other, and that in several cases it is that one which mimics the protected species, while the one that least requires protection never does so, it will afford very strong corroborative evidence that there is a real connexion between the necessity for protection and the phenomenon of mimicry. Now the sexes of insects offer us a test of the nature here indicated, and appear to furnish one of the most conclusive arguments in favour of the theory that the phenomena termed "mimicry" are produced by natural selection.

The comparative importance of the sexes varies much in different classes of animals. In the higher vertebrates, where the number of young produced at a birth is small and the same individuals breed many years in succession, the preservation of both sexes is almost equally important. In all the numerous cases in which the male protects the female and her offspring, or helps to supply them with food, his importance in the economy of nature is proportionately increased, though it is never perhaps quite equal to that of the female. In insects the case is very different; they pair but once in their lives, and the prolonged existence of the male is in most cases quite unnecessary for the continuance of the race. The female, however, must continue to exist long enough to deposit her eggs in a place adapted for the development and growth of the progeny. Hence there is a wide difference in the need for protection in the two sexes; and we should, therefore, expect to find that in some cases the special protection given to the female was in the male less in amount or altogether wanting. The facts entirely confirm this expectation. In the spectre insects (Phasmidæ) it is often the females alone that so strikingly resemble leaves, while the males show only a rude approximation. The male Diadema misippus is a very handsome and conspicuous butterfly, without a sign of protective or imitative colouring, while the female is entirely unlike her partner, and is one of the most wonderful cases of mimicry on record, resembling most accurately the common Danais chrysippus, in whose company it is often found. So in several species of South American Pieris, the males are white and black, of a similar type of colouring to our own "cabbage" butterflies, while the females are rich yellow and buff, spotted and marked so as exactly to resemble species of Heliconidæ with which they associate in the forest. In the Malay archipelago is found a Diadema which had always been considered a male insect on account of its glossy metallic-blue tints, while its companion of sober brown was looked upon as the female. I discovered, however, that the reverse is the case, and that the rich and glossy colours of the female are imitative and protective, since they cause her exactly to resemble the common Euploea midamus of the same regions, a species which has been already mentioned in this essay as mimicked by another butterfly, Papilio paradoxa. I have since named this interesting species Diadema anomala (see the Transactions of the Entomological Society, 1869, p. 285). In this case, and in that of Diadema misippus, there is no difference in the habits of the two sexes, which fly in similar localities; so that the influence of "external conditions" cannot be invoked here as it has been in the case of the South American Pieris pyrrha and allies, where the white males frequent open sunny places, while the Heliconia-like females haunt the shades of the forest.

We may impute to the same general cause (the greater need of protection for the female, owing to her weaker flight, greater exposure to attack, and supreme importance)— the fact of the colours of female insects being so very generally duller and less conspicuous than those of the other sex. And that it is chiefly due to this cause rather than to what Mr. Darwin terms "sexual selection" appears to be shown by the otherwise inexplicable fact, that in the groups which have a protection of any kind independent of concealment, sexual differences of colour are either quite wanting or slightly developed. The Heliconidæ and Danaidæ, protected by a disagreeable flavour, have the females as bright and conspicuous as the males, and very rarely differing at all from them. The stinging Hymenoptera have the two sexes equally well coloured. The Carabidæ, the Coccinellidæ, Chrysomelidæ, and the Telephori have both sexes equally conspicuous, and seldom differing in colours. The brilliant Curculios, which are protected by their hardness, are brilliant in both sexes. Lastly, the glittering Cetoniadæ and Buprestidæ, which seem to be protected by their hard and polished coats, their rapid motions, and peculiar habits, present few sexual differences of colour, while sexual selection has often manifested itself by structural differences, such as horns, spines, or other processes.

Cause of the dull Colours of Female Birds.

The same law manifests itself in Birds. The female while sitting on her eggs requires protection by concealment to a much greater extent than the male; and we accordingly find that in a large majority of the cases in which the male birds are distinguished by unusual brilliancy of plumage, the females are much more obscure, and often remarkably plain-coloured. The exceptions are such as eminently to prove the rule, for in most cases we can see a very good reason for them. In particular, there are a few instances among wading and gallinaceous birds in which the female has decidedly more brilliant colours than the male; but it is a most curious and interesting fact that in most if not all these cases the males sit upon the eggs; so that this exception to the usual rule almost demonstrates that it is because the process of incubation is at once very important and very dangerous, that the protection of obscure colouring is developed. The most striking example is that of the gray phalarope (Phalaropus fulicarius). When in winter plumage, the sexes of this bird are alike in colouration, but in summer the female is much the most conspicuous, having a black head, dark wings, and reddish-brown back, while the male is nearly uniform brown, with dusky spots. Mr. Gould in his "Birds of Great Britain" figures the two sexes in both winter and summer plumage, and remarks on the strange peculiarity of the usual colours of the two sexes being reversed, and also on the still more curious fact that the "male alone sits on the eggs," which are deposited on the bare ground. In another British bird, the dotterell, the female is also larger and more brightly-coloured than the male; and it seems to be proved that the males assist in incubation even if they do not perform it entirely, for Mr. Gould tells us, "that they have been shot with the breast bare of feathers, caused by sitting on the eggs." The small quail-like birds forming the genus Turnix have also generally large and bright-coloured females, and we are told by Mr. Jerdon in his "Birds of India" that "the natives report that during the breeding season the females desert their eggs and associate in flocks while the males are employed in hatching the eggs." It is also an ascertained fact, that the females are more bold and pugnacious than the males. A further confirmation of this view is to be found in the fact (not hitherto noticed) that in a large majority of the cases in which bright colours exist in both sexes incubation takes place in a dark hole or in a dome-shaped nest. Female kingfishers are often equally brilliant with the male, and they build in holes in banks. Bee-eaters, trogons, motmots, and toucans, all build in holes, and in none is there any difference in the sexes, although they are, without exception, showy birds. Parrots build in holes in trees, and in the majority of cases they present no marked sexual difference tending to concealment of the female. Woodpeckers are in the same category, since though the sexes often differ in colour, the female is not generally less conspicuous than the male. Wagtails and titmice build concealed nests, and the females are nearly as gay as their mates. The female of the pretty Australian bird *Pardalotus punctatus*, is very conspicuously spotted on the upper surface, and it builds in a hole in the ground. The gay-coloured hang-nests (Icterinæ) and the equally brilliant tanagers may be well contrasted; for the former, concealed in their covered nests, present little or no sexual difference of colour — while the open-nested tanagers have the females dull-coloured and sometimes with almost protective tints. No doubt there are many individual exceptions to the rule here indicated, because many and various causes have combined to determine both the colouration and the habits of birds. These have no doubt acted and re-acted on each other; and when conditions have changed one of these characters may often have become modified, while the other, though useless, may continue by hereditary descent an apparent exception to what otherwise seems a very general rule. The facts presented by the sexual differences of colour in birds and their mode of nesting, are on the whole in perfect harmony with that law of protective adaptation of colour and form, which appears to have checked to some extent the powerful action of sexual selection, and to have materially influenced the colouring of female birds, as it has undoubtedly done that of female insects.

Use of the gaudy Colours of many Caterpillars.

Since this essay was first published a very curious difficulty has been cleared up by the application of the general principle of protective colouring. Great numbers of caterpillars are so brilliantly marked and coloured as to be very conspicuous even at a considerable distance, and it has been noticed that such caterpillars seldom hide themselves. Other species, however, are green or brown, closely resembling the colours of the substances on which they feed, while others again imitate sticks, and stretch themselves out motionless from a twig so as to look like one of its branches. Now, as caterpillars form so large a part of the food of birds, it was not easy to understand why any of them should have such bright colours and markings as to make them specially visible. Mr. Darwin had put the case to me as a difficulty from another point of view, for he had arrived at the conclusion that brilliant colouration in the animal kingdom is mainly due to sexual selection, and this could not have acted in the case of sexless larvæ. Applying here the analogy of other insects, I reasoned, that since some caterpillars were evidently protected by their imitative colouring, and others by their spiny or hairy bodies, the bright colours of the rest must also be in some way useful to them. I further thought that as some butterflies and moths were greedily eaten by birds while others were distasteful to them, and these latter were mostly of conspicuous colours, so probably these brilliantly coloured caterpillars were distasteful, and therefore never eaten by birds. Distastefulness alone would however be of little service to caterpillars, because their soft and juicy bodies are so delicate, that if seized and afterwards rejected by a bird they would almost certainly be killed. Some constant and easily perceived signal was therefore necessary to serve as a warning to birds never to touch these uneatable kinds, and a very gaudy and conspicuous colouring with the habit of fully exposing themselves to view becomes such a signal, being in strong contrast with the green or brown tints and retiring habits of the eatable kinds. The subject was brought by me before the Entomological Society (see Proceedings, March 4th, 1867), in order that those members having opportunities for making observations might do so in the following summer; and I also wrote a letter to the *Field* newspaper, begging that some of its readers would cooperate in making observations on what insects were rejected by birds, at the same time fully explaining the great interest and scientific importance of the problem. It is a curious example of how few of the country readers of that paper are at all interested in questions of simple natural history, that I only obtained one answer from a gentleman in Cumberland, who gave me some interesting observations on the general dislike and abhorrence of all birds to the "Gooseberry Caterpillar," probably that of the Magpie-moth (Abraxas grossulariata). Neither young pheasants, partridges, nor wild-ducks could be induced to eat it, sparrows and finches never touched it, and all birds to whom he offered it rejected it with evident dread and abhorrence. It will be seen that these observations are confirmed by those of two members of the Entomological Society to whom we are indebted for more detailed information.

In March, 1869, Mr. J. Jenner Weir communicated a valuable series of observations made during many years, but more especially in the two preceding summers, in his aviary, containing the following birds of more or less insectivorous habits:— Robin, Yellow–Hammer, Reed-bunting, Bullfinch, Chaffinch, Crossbill, Thrush, Tree–Pipit, Siskin, and Redpoll. He found that hairy caterpillars were uniformly rejected; five distinct species were quite unnoticed by all his birds, and were allowed to crawl about the aviary for days with impunity. The spiny caterpillars of the Tortoiseshell and Peacock butterflies were equally rejected; but in both these cases Mr. Weir thinks it is the taste, not the hairs or spines, that are disagreeable, because some very young caterpillars of a hairy species were rejected although no hairs were developed, and the smooth pupæ of the above-named butterflies were refused as persistently as the spined larvæ. In these cases, then, both hairs and spines would seem to be mere signs of uneatableness.

His next experiments were with those smooth gaily-coloured caterpillars which never conceal themselves, but on the contrary appear to court observation. Such are those of the Magpie moth (Abraxas grossulariata), whose caterpillar is conspicuously white and black spotted — the Diloba coeruleocephala, whose larvæ is pale yellow with a broad blue or green lateral band — the Cucullia verbasci, whose larvæ is greenish white with yellow bands and black spots, and Anthrocera filipendulæ (the six spot Burnet moth), whose caterpillar is yellow with black spots. These were given to the birds at various times, sometimes mixed with other kinds of larvæ which were greedily eaten, but they were in every case rejected apparently unnoticed, and were left to crawl about till they died.

The next set of observations were on the dull-coloured and protected larvæ, and the results of numerous experiments are thus summarised by Mr. Weir. "All caterpillars whose habits are nocturnal, which are dull coloured, with fleshy bodies and smooth skins, are eaten with the greatest avidity. Every species of green caterpillar is also much relished. All Geometræ, whose larvæ resemble twigs as they stand out from the plant on their anal prolegs, are invariably eaten."

At the same meeting Mr. A. G. Butler, of the British Museum, communicated the results of his observations with lizards, frogs, and spiders, which strikingly corroborate those of Mr. Weir. Three green lizards (Lacerta viridis) which he kept for several years, were very voracious, eating all kinds of food, from a lemon cheesecake to a spider, and devouring flies, caterpillars, and humble bees; yet there were some caterpillars and moths which they would seize only to drop immediately. Among these the principal were the caterpillar of the Magpie moth (Abraxas grossulariata) and the perfect six spot Burnet moth (Anthrocera filipendulæ). These would be first seized but invariably dropped in disgust, and afterwards left unmolested. Subsequently frogs were kept and fed with caterpillars from the garden, but two of these — that of the before-mentioned Magpie moth, and that of the V. moth (Halia wavaria), which is green with conspicuous white or yellow stripes and black spots — were constantly rejected. When these species were first offered, the frogs sprang at them eagerly and licked them into their mouths; no sooner, however, had they done so than they seemed to be aware of the mistake that they had made, and sat with gaping mouths, rolling their tongues about until they had got quit of the nauseous morsels.

With spiders the same thing occurred. These two caterpillars were repeatedly put into the webs both of the geometrical and hunting spiders (Epeira diadema and Lycosa sp.), but in the former case they were cut out and allowed to drop; in the latter, after disappearing in the jaws of their captor down his dark silken funnel, they invariably reappeared, either from below or else taking long strides up the funnel again. Mr. Butler has observed lizards fight with and finally devour humble bees, and a frog sitting on a bed of stone-crop leap up and catch the bees which flew over his head, and swallow them, in utter disregard of their stings. It is evident, therefore, that the possession of a disagreeable taste or odour is a more effectual protection to certain conspicuous caterpillars and moths, than would be even the possession of a sting.

The observations of these two gentlemen supply a very remarkable confirmation of the hypothetical solution of the difficulty which I had given two years before. And as it is generally acknowledged that the best test of the truth and completeness of a theory is the power which it gives us of prevision, we may I think fairly claim this as a case in which the power of prevision has been successfully exerted, and therefore as furnishing a very powerful argument in favour of the truth of the theory of Natural Selection.

Summary.

I have now completed a brief, and necessarily very imperfect, survey of the various ways in which the external form and colouring of animals is adapted to be useful to them, either by concealing them from their enemies or from the creatures they prey upon. It has, I hope, been shown that the subject is one of much interest, both as regard a true comprehension of the place each animal fills in the economy of nature, and the means by which it is enabled to maintain that place; and also as teaching us how important a part is played by the minutest details in the structure of animals, and how complicated and delicate is the equilibrium of the organic world.

My exposition of the subject having been necessarily somewhat lengthy and full of details, it will be as well to recapitulate its main points.

There is a general harmony in nature between the colours of an animal and those of its habitation. Arctic animals are white, desert animals are sand-coloured; dwellers among leaves and grass are green; nocturnal animals are dusky. These colours are not universal, but are very general, and are seldom reversed. Going on a little further, we find birds, reptiles, and insects, so tinted and mottled as exactly to match the rock, or bark, or leaf, or flower, they are accustomed to rest upon,— and thereby effectually concealed. Another step in advance, and we have insects which are formed as well as coloured so as exactly to resemble particular leaves, or sticks, or mossy twigs, or flowers; and in these cases very peculiar habits and instincts come into play to aid in the deception and render the concealment more complete. We now enter upon a new phase of the phenomena, and come to creatures whose colours neither conceal them nor make them like vegetable or mineral substances; on the contrary, they are conspicuous enough, but they completely resemble some other creature of a quite different group, while they differ much in outward appearance from those with which all essential parts of their organization show them to be really closely allied. They appear like actors or masqueraders dressed up and painted for amusement, or like swindlers endeavouring to pass themselves off for well-known and respectable members of society. What is the meaning of this strange travestie? Does Nature descend to imposture or masquerade? We answer, she does not. Her principles are too severe. There is a use in every detail of her handiwork. The resemblance of one animal to another is of exactly the same essential nature as the resemblance to a leaf, or to bark, or to desert sand, and answers exactly the same purpose. In the one case the enemy will not attack the leaf or the bark, and so the disguise is a safeguard; in the other case it is found that for various reasons the creature resembled is passed over, and not attacked by the usual enemies of its order, and thus the creature that resembles it has an equally effectual safeguard. We are plainly shown that the disguise is of the same nature in the two cases, by the occurrence in the same group of one species resembling a vegetable substance, while another resembles a living animal of another group; and we know that the creatures resembled, possess an immunity from attack, by their being always very abundant, by their being conspicuous and not concealing themselves, and by their having generally no visible means of escape from their enemies; while, at the same time, the particular quality that makes them disliked is often very clear, such as a nasty taste or an indigestible hardness. Further examination reveals the fact that, in several cases of both kinds of disguise, it is the female only that is thus disguised; and as it can be shown that the female needs protection much more than the male, and that her preservation for a much longer period is absolutely necessary for the continuance of the race, we have an additional indication that the resemblance is in all cases subservient to a great purpose — the preservation of the species.

In endeavouring to explain these phenomena as having been brought about by variation and natural selection, we start with the fact that white varieties frequently occur, and when protected from enemies show no incapacity for continued existence and increase. We know, further, that varieties of many other tints occasionally occur; and as "the survival of the fittest" must inevitably weed out those whose colours are prejudicial and preserve those whose colours are a safeguard, we require no other mode of accounting for the protective tints of arctic and desert animals. But this being granted, there is such a perfectly continuous and graduated series of examples of every kind of protective imitation, up to the most wonderful cases of what is termed "mimicry," that we can find no place at which to draw the line, and say,— so far variation and natural selection will account for the phenomena, but for all the rest we require a more potent cause. The counter theories that have been proposed, that of the "special creation" of each imitative form, that of the action of "similar conditions of existence" for some of the cases, and of the laws of "hereditary descent and the reversion to ancestral forms" for others,— have all been shown to be beset with difficulties, and the two latter to be directly contradicted by some of the most constant and most remarkable of the facts to be accounted for.

General deductions as to Colour in Nature.

The important part that "protective resemblance" has played in determining the colours and markings of many groups of animals, will enable us to understand the meaning of one of the most striking facts in nature, the uniformity in the colours of the vegetable as compared with the wonderful diversity of the animal world. There appears no good reason why trees and shrubs should not have been adorned with as many varied hues and as strikingly designed patterns as birds and butterflies, since the gay colours of flowers show that there is no incapacity in vegetable tissues to exhibit them. But even flowers themselves present us with none of those wonderful designs, those complicated arrangements of stripes and dots and patches of colour, that harmonious blending of hues in lines and bands and shaded spots, which are so general a feature in insects. It is the opinion of Mr. Darwin that we owe much of the beauty of flowers to the necessity of attracting insects to aid in their fertilisation, and that much of the development of colour in the animal world is due to "sexual selection," colour being universally attractive, and thus leading to its propagation and increase; but while fully admitting this, it will be evident from the facts and arguments here brought forward, that very much of the *variety* both of colour and markings among animals is due to the supreme importance of concealment, and thus the various tints of minerals and vegetables have been directly reproduced in the animal kingdom, and again and again modified as more special protection became necessary. We shall thus have two causes for the development of colour in the animal world, and shall be better enabled to understand how, by their combined and separate action, the immense

Mimicry Among Animals

variety we now behold has been produced. Both causes, however, will come under the general law of "Utility," the advocacy of which, in its broadest sense, we owe almost entirely to Mr. Darwin. A more accurate knowledge of the varied phenomena connected with this subject may not improbably give us some information both as to the senses and the mental faculties of the lower animals. For it is evident that if colours which please us also attract them, and if the various disguises which have been here enumerated are equally deceptive to them as to ourselves, then both their powers of vision and their faculties of perception and emotion, must be essentially of the same nature as our own — a fact of high philosophical importance in the study of our own nature and our true relations to the lower animals.

Conclusion.

Although such a variety of interesting facts have been already accumulated, the subject we have been discussing is one of which comparatively little is really known. The natural history of the tropics has never yet been studied on the spot with a full appreciation of "what to observe" in this matter. The varied ways in which the colouring and form of animals serve for their protection, their strange disguises as vegetable or mineral substances, their wonderful mimicry of

96

other beings, offer an almost unworked and inexhaustible field of discovery for the zoologist, and will assuredly throw much light on the laws and conditions which have resulted in the wonderful variety of colour, shade, and marking which constitutes one of the most pleasing characteristics of the animal world, but the immediate causes of which it has hitherto been most difficult to explain.

If I have succeeded in showing that in this wide and picturesque domain of nature, results which have hitherto been supposed to depend either upon those incalculable combinations of laws which we term chance or upon the direct volition of the Creator, are really due to the action of comparatively well-known and simple causes, I shall have attained my present purpose, which has been to extend the interest so generally felt in the more striking facts of natural history to a large class of curious but much neglected details; and to further, in however slight a degree, our knowledge of the subjection of the phenomena of life to the "Reign of Law."